I0005339

E-Doc-S

E-Doc-S

◆

Hail the Freelance Renaissance

Sean Kinn

iUniverse, Inc.
New York Lincoln Shanghai

E-Doc-S
Hail the Freelance Renaissance

iUniverse books may be ordered through booksellers or by contacting:

iUniverse
2021 Pine Lake Road, Suite 100
Lincoln, NE 68512
www.iuniverse.com
1-800-Authors (1-800-288-4677)

Because of the dynamic nature of the Internet, any Web addresses
or links contained in this book may have changed
since publication and may no longer be valid.

The views expressed in this work are solely those of the author and do
not necessarily reflect the views of the publisher, and the publisher
hereby disclaims any responsibility for them.

ISBN: 978-0-595-45135-7 (pbk)
ISBN: 978-0-595-89447-5 (ebk)

Printed in the United States of America

To my sons, Niko and Dano, who suffered the long absences while I served my country.

Nothing in the world can take the place of persistence. Talent will not—nothing is more common than unsuccessful men with talent. Genius will not—unrewarded genius is almost a proverb. Education will not—the world is full of educated derelicts. Persistence and determination alone are omnipotent.

—Calvin Coolidge, 1872–1933

Contents

Preface

This book describes selling edocuments online. To that end, I added as much related information as I thought necessary to support my thought process.

While writing the book, I examined the math involved with edocument publishing as well as the ease with which a sales platform may be set up, and I have convinced myself this new type of writing is the biggest breakthrough for nonfiction writers since the creation of freelance query letters and book proposals. To rephrase what I just said, you no longer *need* query letters and book proposals if you sell text directly to readers by way of edocuments.

I've watched edocument sales since at least 2001 and the best sales potential indicator I have seen to date is the relatively recent revision and expansion of Amazon.com's edocument sales section. When Amazon began selling short fiction—Amazon Shorts—I knew I was witnessing something special. (This book isn't about fiction—it's about nonfiction edocuments—the Amazon Shorts thing is still cool.)

Having said all of the above, the Internet has facilitated a number of changes in the publishing world, so is it any wonder writers are selling text online for what essentially amounts to hundreds of dollars a word? Revenue arrives incrementally over time, but it's a fact. Interestingly, the writers who have managed this feat did so simply by reorganizing the structure of their writing.

Another attractive feature of the edocument purchase process is that it's possible to search for nonfiction content online. Whereas, be my guest and visit a physical world used bookstore to search through old magazines for needed subject matter.

If I were to make a prediction I would say we are on the verge of a freelance writer renaissance and like a lot of things these days—it just happened—via innovation, time, and efficiency inherent to computers and the Internet. Put another way, a few highly creative, Web savvy writers connected the dots that exist between electronic documents and credit card sales and began accepting money directly from readers for their words—and the rest is history.

Acknowledgments

I would like to acknowledge my best friend the world over, JTL. Our early morning, overseas discussions are often the most uplifting thing in my life. I would like to thank my mentors—they range from freelance writers who are much better wordsmiths than I could ever be—to publishers, editors, reporters, officers, and gentlemen. They know who they are. I would also like to thank my dog, Castro, who was forced to listen to my gibberish during editing.

1

Edocster

I'd buy that for a dollar!

—*Bixby Snyder,* **Robocop***, 1987*

The concept of exchanging digital material in piece-meal fashion is not new. The music world began separating compact discs into individual song downloads in the late 1990s. I'm not sure what took the writing world so long to do the same.

An edocument is essentially any saleable digital text file. The text files are generally created in word processor software. Graphics may be added to the text files if desired. The files are converted to an online readable format, advertised on the Internet and—presto—sold directly to readers. As simplistic as it sounds—that's how it works.

Edocuments eliminate three writer's tools that are normally an essential part of being published in the nonfiction world:

- The freelance query letter.

- The nonfiction book proposal.

- The nonfiction book table of contents.

With edocuments, you no longer need to write a letter or send an e-mail to an editor to ask for permission to begin writing an article. You don't need to spend weeks or months outlining a book proposal. You no longer need to toil with a list of contents because edocuments may be sold separately with no connection to a larger book organization.

Alternately, you can still have a table of contents. If I decide to expand and sell the case studies contained in this book as separate edocuments, I could design a Web page to hyperlink to individual edocument sales pages. Additionally, there's not much difference between a large edocument and a small ebook these days. As in, a large edocument could still include a self-contained list of contents or a hyperlinked index.

2

Amazing

o o o o o o o o o o o o o o o o o o o o o o o o o o o

He can compress the most words into the smallest ideas better than any man I ever met.

—Abraham Lincoln, 1809-65

To get a handle on where the edocument market is at today review Amazon's edocument section:

• Go to "Amazon Shorts" in their main product categories.

• Find the "Buy eDocs" link; usually in the bottom left margin of the page.

• From there you'll find a growing collection of edocuments sold at Amazon by large-volume-producer publishing entities.

Here's an abbreviated list of Amazon's edocument nonfiction content (the list may change after this book goes to print):

- Biographies and Memoirs

- Business and Investing

- Computers and Internet

- Cooking, Food, and Wine

- Health, Mind, and Body

- History

- Home and Garden

- Reference

- Religion and Spirituality

- Romance

- Science and Technology

- Sports

- Travel

Here's an abbreviated list of large-volume-producer publishing entities selling the above edocument content at Amazon. You may have heard of some of these publishers:

- BrownHerron Publishing
- Consumer Reports
- Credit Suisse First Boston
- DevX
- Gale Group
- Gobi International
- Harvard Business Review
- International Business Publications
- MIT Technology Review
- McKinsey & Company
- Motley Fool
- Princeton University Press
- Salon
- Standish Group
- TIME Magazine
- Tect Ltd.
- The Heritage Foundation
- The Wall Street Transcript

- Zacks Investment Research

- Brandon-hall

If you're still not impressed, here's a separate list of edocument content I discovered while researching for this book:

- Bodybuilding tips

- Business and family tip lists

- College study guides

- Federal job tips

- Medical journals

- Organization tips

- Professional writing how-to

- Publicity tip lists

- Sales letters

- Sales-training products

The most recent online edocument publisher I ran across was a federal job assistance Web site that sells sample KSAs—sample lists of employee knowledge, skills, and abilities—used to apply for certain federal jobs.

At this point, I'm not suggesting freelance writers should halt their query efforts and suddenly begin writing everything on spec to post articles online. My shift toward edocument publishing came about gradually while testing various forms of freelance writing over the years. With time I came to believe query letters and nonfiction book proposals are time-consuming and an inherently slow way to communicate about a particular writing project. The edocuments I'm developing at present are ideas that magazine and book publishers may want to see, but I'll never know because I'm taking them directly to the Web to provide readers with an opportunity to buy from the writer.

Because I'm putting so much effort into this type of work, if a publisher is interested in reprinting something I have written, the publisher will have to initiate contact. I don't consider this arrogance. It's merely a lack of time on my part.

3

Consumer Convenience

o o o o o o o o o o o o o o o o o o o o o o o o o o o

There it is, the world's first convenience store.

—Apu, **The Simpsons: Homer and Apu,** 1994

Edocument publishing may be changing nonfiction text sales in such a way that readers now prefer the ease of an online purchase over driving or walking to the local bookstore or grocery. There are tons of paper nonfiction books, consumer magazines, and newspapers available in physical world stores and that's not going to change anytime soon. However, anyone familiar with human nature knows that people are often easily overcome by the simplest of things—convenience—for example.

If you combine Internet search with convenience, you may be able to see why some readers are spending more time looking for information online, finding it, and buying it: The edocument publishing model obviously benefits readers.

I believe a slow-motion scramble to author edocument content is happening as we speak and will continue to grow as writers are told they can sell directly to readers, and readers discover they can purchase directly from writers.

4

New Freelance Math

○ ○

Defendit numerus: There is safety in numbers.

—*J. R. Newman (ed.),*
The World of Mathematics, *1956*

The most fascinating part of the new edocument publishing equation is the potential math. Rhetorical question: "One edocument sold online numerous times equals how much money for the writer?"
    Answer:

*   Revenue generated by edocument sales—one edocument sold a number of times—is *proportional* to the number of copies sold.

*   The revenue builds over a period of time and becomes *cumulative* in one person's coffers (the writer's).

- If you add the aforementioned *consumer convenience*—ubiquitous accessibility, worldwide availability, and user-friendly purchase—and *perpetual sales* to the formula—you may be able to envision an automated windfall.

The math equals a combination of freelance writing and traditional self-publishing.

- The windfall difference between the traditional self-publishing model and an automated edocument sales platform can be defined by stating that you normally must use a portion of your profits to reprint books—whereas, with edocuments—you do not.

- The big change in so far as freelance revenue—with reference to old freelance math—is that instead of receiving a one-time freelance fee—a few hundred dollars for the sale of a thousand-word article to a magazine, for example—the writer receives *royalties*.

The royalties-versus-fee marvel may be explained by noting that as an edocument publisher you fill in for or replace the following publishing roles and functions:

- Author.

- Cover designer; but only if you think you need a cover for your edocument.

- Content editor.

- Copy editor.

- Typesetter; computers and software accomplish this for you.

- Printer, paper, and ink; although there is nothing to print unless the reader opts to print something.

- Binder; the need for this is eliminated.

- Publisher; this is just a description of the entire process.

- Sales representative; this position is eliminated.

- Distributor; with assistance provided by your edocument file Web host.

- Warehouse and bookstore owner; figuratively, of course.

- Reviewer.

- Advertising agency.

- Bookkeeper.

- Secretary; this is actually the toughest management hat to wear because it's you doing everything.

As you can see, many of the various processes are automated and you don't actually perform the management or publishing function. Many of the traditional publishing house positions are "eliminated" by computer and Internet efficiency. What we're actually examining is an entire new arrangement of publishing *economics*.

With edocument sales, the author retains the majority of revenue that may be generated and there are even ways to project royalties. In theory, specific math used to determine total revenue could be figured like so:

- Author a small collection of edocuments—various subject matter—say, ten different edocuments—and price them at $5 apiece.

- Sell a thousand total edocuments—one hundred each of the ten different edocuments—over a period of thirty days.

- Could you subsist as a freelance writer on $5,000 a month?

- If you sold a thousand of each of the ten different edocuments over a period of thirty days could you subsist on $50,000 a month?

- If you thought your content was worth more than $5 a copy, could you charge $100 or $1,000 for an edocument?

In a credible effort to capture the entire Internet's large-volume edocument producers, Amazon has been selling technology and market research edocuments for thousands of dollars apiece for years ("edocuments" that are larger than some ebooks). Go to "Amazon Shorts," "Buy eDocs" (bottom left), "Business & Investing" and then sort by "Price: High to Low."

Since we're speaking of a fluid market I won't name specific edocuments, but I will say: "Go, whoever has a need for four-figure edocuments!"

Does that mean it's possible for freelance writers to sell edocuments for $1,000 apiece and generate $10 million? In a theoretical sense—maybe—but I believe generic math is more indicative of potential and that writers should concentrate on the reality of what they're trying to accomplish. If you review the E-Doc-S Web site http://www.e-doc-s.info that supports this book, my edocuments are either free, 1¢ apiece for test purchases, or appropriately priced if compared to a similar amount of text in a paper nonfiction book (e.g., one chapter). I've priced my edocuments in this manner to preclude refunds.

You can also figure edocument math using the normal freelance writer per-word method:

- Write an edocument that contains five thousand words—any topic—and price it at $5.

- Sell a thousand copies over a period of thirty days.

- As a freelance writer you just wrote for $1 a word.

- A thousand copies multiplied times $5 equals $5,000.

• $5,000 divided by five thousand words equals $1 a word.

The new edocument math appears to have been specifically designed for freelance writers. If you're familiar with traditional freelance writing, you may know how long it takes to work up to a dollar a word. Combine that thought with the fact that edocument revenue increases exponentially based on the amount of effort you put into your work.

To recap this section, did I communicate to you that you will be pulling in tens of thousands of dollars a month after creating a freelance edocument sales platform? No—I have merely addressed revenue potential—that is all.

I do believe there may be more freelance revenue available with edocument sales than that normally associated with traditional freelance writing.

5

Selling Edocuments Online

○ ○

A recent government publication on the marketing of cabbage contains, according to one report, 26,941 words. It is noteworthy in this regard that the Gettysburg Address contains a mere 279 words while the Lord's Prayer comprises but 67.

—Norman R. Augustine, 1935–

There are many ways to sell edocuments online. Before you begin reading the edocument case studies that follow this section, note that you may become a Webmaster of sorts if you design your own sales platform. If you go it alone, you will at a minimum cut and paste HTML provided to you by an edocument Web host. Thankfully, our Friends the Code Writers continue to develop new and exciting ways for writers to work with the Web. I'm semi HTML illiterate, but I get along just fine designing Web pages as long as I do it my way.

If you grew up writing or editing HTML you may laugh at someone not interested in learning this particular type of code (I'm sure it's easy). But you have to realize that I use computers and the Internet as writing tools, and nothing more. My computer experience dates to Wang word processors and Commodore Colts. If you're a writer who has never heard of or used a "C prompt" you may be a relatively young writer. I'm a 40-something writer who enjoys watching and waiting while other people develop new tricks.

The simplest way to explain how to create an edocument goes like this:

• Create a text file.

• Convert the file to a Web-readable format. I use PDF.

• Market it.

• Sell it from a Web-based sales platform, or have a Web entity sell it for you.

It's fairly easy to see standardized edocument formats at this time, but that could change. I use Microsoft Publisher 2007 to create text files because I think Publisher produces tastefully designed edocuments. It hasn't always been possible to convert Publisher to PDF, but it is now and the 2007 version makes it a breeze. I use Microsoft Excel to create charts to paste into edocuments. I can see potential to expand my edocuments into PowerPoint slide shows and other media formats for use in classroom environs. I've used Adobe Acrobat Standard for years to edit PDF files. For me it was worth the money to invest in these software programs while testing various edocument publishing platforms.

If you're on a shoestring budget there are a number of free word processor options available these days. Google recently began offering free word processing and PDF conversion software at Google Docs & Spreadsheets http://docs.google.com. There are many other versions of the same available on the Internet.

If I were to predict a future for edocument sales in general, any entity that perfects user-friendly debit and credit cards sales will succeed. The less effort to make a purchase, the more inclined readers will be to pay money for text. Amazon has this down pat with their 1-Click® system. Google Checkout offers a similarly simple checkout system. As ubiquitous as Amazon's online sales platform is, I've also observed Google's Checkout icon popping up on more and more online store pages of late.

6

BrownHerron Publishing (Amazon)

○ ○

Patience and perseverance have a magical effect before which difficulties disappear and obstacles vanish.

—John Quincy Adams, 1767–1848

I wouldn't recommend Googling as a sole means to conduct research while writing a book, but it definitely helps.

If you Google the terms "BrownHerron edoc" and click on "BrownHerron Publishing FAQ" you will arrive at Brown-Herron Publishing's "Frequently Asked Questions By Authors." Alternately, go to http://www.brownherron.com and review their entire Web site.

Long-story-short, BrownHerron's editors work with writers to place edocuments into Amazon's mammoth online sales machine. Join BrownHerron's fee-based alliance, meet their editorial standards and begin authoring edocuments.

As a publishing entity, they are singularly good at what they do—well worth the fees charged for their service—if I could gush about what they're doing for the edocument market right now, I would.

7

Google Docs, E-Junkie, and Google Checkout

○ ○

All animals are equal but some animals are more equal than others.

—Eric Arthur Blair, otherwise known as "George Orwell," 1903-50

This is my current preferred edocument sales method, a true do-it-yourselfer technique:

- Go to Google Docs http://docs.google.com.

- Create text. For example, generate a list of ten Smartest Things you've ever done in your life. Top of the list may be "Created a Freelance Edocument Online Sales Platform."

- Save your new Google Doc, then re-save it as a Google PDF.

- Download a free copy of Adobe Reader software to enable you to read your new edocument http://www.adobe.com, unless you already have it on your computer (most do).

- Establish a Google Checkout account http://checkout.google.com.

- Set up an E-Junkie account http://www.e-junkie.com. E-Junkie is a fee-based hosting service that allows you to store edocuments for online sales. They provide the HTML button code that completes an edocument sale from your site through E-Junkie to Google Checkout. They offer other sales methods as well; PayPal http://www.paypal.com, 2Checkout http://www.2checkout.com, ClickBank http://www.clickbank.com, and so forth.

- Build a Web site to sell your edocuments. At present I recommend Microsoft Live Web Basics http://officelive.microsoft.com to build a basic site for edocument sales, but you can also use any free or fee-based Web design software. Live Web is great for amateur Web designers—like me—and pages appear professionally designed. The key to your Web platform is low monthly overhead to increase total revenue. Free Web design packages are obviously better than fee-based design packages.

- Once your site is operational, upload edocuments as files to E-Junkie, paste Google Checkout "Buy

Now" buttons on Web pages and you have the potential to sell directly to readers.

If the process sounds complicated, it may be for some writers. If not, pat yourself on the back for your management skills as you assume the roles of author, editor, graphic designer, typesetter, publisher, distribu- tor, kiosk owner and advertising agency.

During set-up and before selling edocuments, you will most likely have to tweak your E-Junkie/Google sales plat- form to ensure it is fully functional. The best way to get started is to go to the "Functional Sales Platform" link at E- Doc-S http://www.e-doc-s.info, observe how an actual edocument sale functions online, and then work your way back to your Web site. In other words, make a test pur- chase—costs 1¢—from one of the sample edocuments at E-Doc-S—or *almos*t make a test purchase (don't com- plete the purchase)—or download a free edocument—to observe the various processes—then establish a com- pleted edocument sale as your eventual goal at your Web site.

Obviously you have to build your Web site from the ground up, inserting text and other content as you build the pages, but you should understand how the edocument purchase process works before you begin.

More details are available at our Gather location http://e- doc-s.gather.com and at E-Doc-S http://www.e-doc-s.info as you work your way through your E-Junkie and Google Checkout account settings.

Also feel free to work with E-Junkie representatives during set-up. E-Junkie reps can answer questions and may conduct 1¢ test purchases for your edocuments once they're online.

8

Yahoo!, PayPal, and Google Page Creator

o o o o o o o o o o o o o o o o o o o o o o o o o o o

My name's William Forrester. (Points to "writers' wall of fame" pictures.) I'm that one up there.

—*Forrester,* **Finding Forrester,** *2000*

I designed a functional Yahoo! SiteBuilder <u>http://sitebuilder.yahoo.com</u> edocument sales platform approximately two years ago and made an inadvertent sale while testing the platform. A peer ran across my test site while Googling my name and purchased an edocument to see what I "was up to."

The point for that anecdotal information is that the Yahoo! method works and works especially well if you are intimately familiar with Yahoo! SiteBuilder and Web Hosting. Google Page Creator <u>http://pages.google.com</u> online software is similar and free as of this writing (Yahoo! version is fee-based). Both are user-friendly.

With Yahoo!, the edocument sales platform functions within the SiteBuilder Web design software. In other words, edocument files are stored on Yahoo! servers and you acquire PayPal "Buy Now" HTML buttons from within Yahoo!'s Small Business Web site once you begin paying for the service.

With Page Creator you can use E-Junkie <u>http://www.e-junkie.com</u> HTML buttons, which include Google Checkout, PayPal, 2Checkout, or ClickBank "Buy Now" button code.

The Yahoo! and Google edocument sales platforms essentially function in the same manner:

- Create text and save it as PDF.

- Upload PDF edocuments to Yahoo! SiteBuilder or Google Page Creator depending on which you use.

- Apply PayPal or E-Junkie "Buy Now" buttons.

A link to a functional Google Page Creator edocument sales platform may be found at our Gather discussion area <u>http://e-doc-s.gather.com</u> and at E-Doc-S <u>http://www.e-doc-s.info</u>.

9

Lulu

○ ○

Should not the Society of Indexers be known as Indexers, Society of, The?

—Keith Waterhouse, 1929–

Lulu's mainstay is their ebook sales program http://www.lulu.com. However, if anyone can accurately describe the difference between a small ebook and a large edocument, please feel free to do so at our Gather location. Meanwhile, it's easy to sell edocuments with Lulu's ebook program:

• Create text and save as PDF. Lulu also provides a PDF conversion service, if you need it.

• Enter information about your edocument to allow customers to search for and find it.

• Upload a cover file or select a free cover image from Lulu.

- Select your price. Lulu charges a percentage fee per sale.

- Select your copyright license.

- Lulu offers a stable linking system which allows you to paste static links into your sales Web site or wherever you decide to promote your work.

A link to a sample Lulu edocument may be found at our Gather discussion area http://e-doc-s.gather.com and at E-Doc-S http://www.e-doc-s.info.

10

Beaucoup Other Methods

o o o o o o o o o o o o o o o o o o o o o o o o o o
*Science becomes dangerous only when
it imagines that it has reached its goal.*
> —George Bernard Shaw, 1856–1950

Hopefully, some of the preceding case studies have illustrated edocument sales platforms are an emergent combination of technologies. With the different Web design packages available these days there are probably tens if not hundreds of ways to combine Web design and edocument hosting with debit and credit card sales. Similarly, in the last few years I have observed a growing number of Web sites open specifically to assist freelance digital product sales.

In addition to the methods described in the case studies, you should also scrutinize other writers' sales techniques. Take a look at the "Edocument Writer" links at our Gather location http://e-doc-s.gather.com and at E-Doc-S http://www.e-doc-s.info. If you happen to see a sales method you are familiar with or have used in the past, by all means, feel free to use that method to build your platform. In the spirit of this book, please also tell the Gather group what you are doing to assist with their efforts.

If I haven't said it often enough, if you strive for one-click simplicity your sales platform will sell more edocuments.

11

Proposed Edocument Content

∘ ∘

Thars gold in them thar hills!

—Mark Twain (Samuel Langhorne Clemens), 1835–1910 (many people believe Yosemite Sam coined this phrase; not true)

Although I can see the direction my niche writing is headed, at this point it would be rash for me to recommend specific edocument content to other writers. Frankly, I think we're looking at a lot of unknowns. Can anyone name the first paper magazine and explain how content branched out from there?

Regardless, unless a writer already has a huge PDF collection of personally copyrighted material at her or his disposal, it may take a while for the writer to build enough edocuments for online sales.

Amazon's edocument section may help you determine which direction to take your area of nonfiction expertise. The key to developing content lies in an understanding of the term "freelance." If you don't already know, the difference between a freelance writer and a reporter is essentially that the freelancer gets to sit around in underpants at the house, dreaming up article ideas, mailing the ideas in query letter format to editors, subsequently receiving rejection notices that may be framed and placed on the wall. I have a big wall for mine. Whereas, staff writers are tied to desks and laptops chomping fingernails while waiting to be assigned articles by an editor. I've done both types of work and I like both.

A general note about my experience with the editorial process is that I have never had a run-in with an editor. I don't have a poor attitude toward the publishing industry. I accept criticism well and very much enjoy reading something I have written that has been well edited.

I organize and write my words easily enough. I keep a running table of contents in my head for anything I write—book or article—until I'm sure I can fill book chapters or article sections. When I'm complete with a final draft, I let my words rest for a minimum of three days and then read the text aloud—verbatim—*in toto*. After which, missing thought process, clunky verbiage and large numbers of typographical errors may stick out like a sore thumb. Depending on the number of times I repeat the process, I may end up with a very small number of typos and my text communicates what it's supposed to communicate. There will always be typographical errors in published text. The trick is to minimize the number of errors.

This read-aloud system basically replicates a professional copy-edit process as if a fresh set of eyes were examining your work. I realize WYSIWYG—what-you-see-is-what-you-get from your computer screen—is very close to a printed piece of paper these days, but I tend to have fewer typos if I print out a copy of my work to read aloud versus attempting the same on the screen especially for the final read-aloud.

As an edocument publisher, you eliminate the proposal go-ahead portion of the traditional freelance publishing process by writing something and posting it to the Web—keeping in mind there are a few billion available readers—simplistic to state, but true:

• If you are a traditional freelance writer and you have ideas that have been rejected—articles ideas or book proposals—these are items you should examine for edocument sales potential.

• If you have articles you have written on spec in the past, these should be looked at to see if it would be possible to convert them into PDFs to immediately start building your edocument collection.

• If you are an expert in your field you are a prime candidate to create and sell knowledge to readers. Especially so if you have nonfiction book(s) out in print.

In general terms you must write what you know and you should be somewhat credentialed. I wrote my first feature magazine article seven years before I received a bachelor's in communication studies. University degrees do not necessarily define "credentials." At a minimum, you must have a content angle, preferably a niche subject area where there is little or no competition, and readers must *need* your words. Or *want* them terribly; which is essentially the same thing.

In my case, I've held off on querying certain topics for the last few years for reasons I cannot explain other than to state that it may have been instinct. No big clairvoyance on my part, but it is possible I have become somewhat stingy with my writing ideas. Maybe I've been writing for long enough to know a good article idea when I see one all by my lonesome. It's a logical evolution for writers to become editors after lengthy periods of keyboard pounding.

12

Niche Writing

o o o o o o o o o o o o o o o o o o o o o o o o o o

Irony is a disciplinarian feared only by those who do not know it, but cherished by those who do.

—*Søren Aabye Kierkegaard, 1813-55*

The following are samples of where I plan to take my niche edocument writing. This is not an all-encompassing list of proposed personal edocument content. I don't get writers block, but it would still be amateurish for me to give away my entire list of ideas.

The purpose for detailing this list of ideas is to illustrate to other readers how easy it is to generate content:

• "Communication studies" case studies are easy to create. I enjoy observing other writers beat Catch 22s, especially when it involves new or innovative forms of freelance writing. I dislike negative responses and enjoy change. Enough said on that topic.

- I would like to pass on my military management experience to pure civilians (those who never had the opportunity to serve): Dumbed-down spreadsheets used to manage groups of people and tasks; standing operating procedures that may save people from misplacing their car keys (for the rest of their lives); the military's version of backward planning, and so forth.

- I'm working on a catalog of adventure travel articles. I'm fairly well traveled and a number of the countries I've visited over the years were dangerous to the extreme. Some might consider hobby adventure travel to be odd, but that's what it is for me, and I've decided it's something I will do for the rest of my days. I'm a former anti-terrorism officer and enjoy pointing out the obvious to other world trekkers before they go. In addition to my travels, I've been stuck in a foreign country for the last quarter century, so why not write about it.

- That's my short list. In general terms, everything I write from this point forward—blog entry, Web page, or nonfiction book chapter—will become an edocument.

If it's not obvious yet, not all writers will be cut out for edocument publishing. If you're a nonfiction author who has already succeeded—if you have books out with a major publishing house and are doing well (better than subsisting)—you may not want to delve into something like this.

If you're a mainstream nonfiction writer and famous, albeit starving somewhat, maybe you will want to take the plunge.

Would edocument publishing be something for writers who are new to freelancing? To answer my question, I wonder if there are edocument writers who began their careers without realizing they bypassed the query letter and book proposal?

13

Switch to Edocument Publishing

○ ○

If a composer could say what he had to say in words he would not bother trying to say it in music.

—Gustav Mahler, 1860–1911

I professionally study various forms of publishing. Thus, I have a large writer's reference library. After buying a book, I may flip through it and file it for future reference. I also buy books specifically to learn—absorb, highlight, and bookmark—and tote around until dog-eared.

The last time I bought a nonfiction book it was because I needed a piece of information to complete a project (to confirm thought process). Other readers may have purchased the same book and required everything from front cover to back matter. To reiterate,

I purchased the book knowing full well I was paying for unneeded chapters.

To make the "edocster" reference once more, the *needed* facet of my last nonfiction book purchase is about the same as buying a music CD or album for a specific song—a *hit* song heard on the radio—which has been standard practice for music lovers for years.

The case studies that follow this section include lessons-learned that caused me to discard or pause a particular type of freelance writing. I've also included notes in some places concerning potential money that may be made from the type of writing.

14

Nonfiction Book Self-Publishing

○ ○

I love being a writer. What I can't stand is the paperwork.

—Peter De Vries, 1910-93

It took me two years of keyboard pounding to self-publish my first nonfiction book in the late 1990s. I burned up two notebook computers and a laser printer and mastered Microsoft Publisher 2.0 software along the way. I wrote and designed the book cover-to-index taking desktop-published pages to a local printer to have it printed into a galley and bound into a book.

About the same time the second edition came out, a study program I described in the book switched from paper application forms to online application forms and that was it: Zap. Dated book. I halted distribution and chalked up the non-event to experience.

I was still happy. My small press efforts had essentially created an urban myth within military circles. I had proven self-publishing windfalls exist. If I hadn't been on active duty and in the field for an average of nine months out of any given year I'm sure I could have worked around the Internet efficiency issue and continued this type of writing for the duration.

First edition sales were brisk enough to cause a distributor to reconsider a 24/7 toll-free book-order telephone line. An article about the book ran in a military newspaper that enjoyed circulation at overseas U.S. military bases and in the United States. In the weeks after the story ran the mom-and-pop distributor began receiving orders in the middle of the night from far away places like Germany and Japan. This, in turn, turned out to be a positive development because when the distributor was no longer able to answer the phone (lack of sleep), soldiers started calling the newspaper to attempt to order the book. Soon thereafter a marketing rep from the paper contacted me to offer to advertise the book for free in return for a percentage of retail. I said "yes," of course.

I could generate a long list of reasons why this experience eventually pushed me toward online edocument publishing. The foremost reason would be the amount of time I spent attempting mass distribution. To summarize:

- Every writer wants mass distribution; the more, the merrier. I pushed both editions of my book into overseas U.S. military bookstores and libraries and was able to physically watch at least one book a month disappear off bookstore shelves in Germany. I used Amazon's Advantage program to crack the code on online sales, and would have eventually managed stateside military bookstore sales, but the Internet efficiency issue occurred and I switched to military correspondent work and freelance articles for the short-term.

- Based on the word count for the two book editions—both around 35,000 words—I would estimate this type of writing at about $1 a word when balanced against an industry-standard first printing, cover price and total sales. The time and elbow grease factors—phone calls, printed invoices, book shipping and so forth—disappear completely with the edocument publishing process.

- A small amount of edocument trivia is that I'm converting the original Microsoft Publisher nonfiction book page files into PDF edocuments to post online as published clips. Later I plan to update the original book project in edocument format.

15

Freelance Articles

○ ○

The free-lance writer is a man who is paid per piece or per word or perhaps.

—Robert Benchley, 1889–1945

Being a traditional freelance writer beats being a staff writer hands-down because you get to set your schedule. When I'm freelancing full-time, I get up very early in the morning, and usually exhaust my creativity by noon. A quick nap and then the rest of the day is mine. My eventual goal is to work fewer than thirty hours a week when I'm working.

Thus far the most I've written for is two dollars a word. After developing my version of the query process way back when, it became a matter of responding to editors with appropriate text. My only problem has been sending too many queries.

The ability to sell text directly to readers without having to query or propose would seem to be a win-

win situation for those of us who hate neckties. I may spend five hours of any given work week crafting query letters; however, if I could use that time for other things, I would be willing to do so.

I don't mind the query process although it is slow which means it is inefficient. Every writer enjoys receiving the go-ahead from an editor for a piece of writing, but if I had a house payment to make I'm not sure if freelancing in the traditional manner would be consistent enough to make payments.

16

Staff Writer

If one morning I walked on top of the water across the Potomac River, the headline that afternoon would read: "President Can't Swim."

—Lyndon B. Johnson

Being a staff writer or reporter can be a rewarding job. I'm sure there are people who enjoy the routine of showing up every day to receive writing assignments. There is definitely much to be said for the security and benefits associated with a day job. I know newspapers and magazines are group efforts that are more power-ful as a whole than any given individual writer's agenda could ever be.

Still, the questions I would pose to reporters are:

• What are your employment contingency plans?

- How secure are your jobs in today's modern writing world?

- Have you ever thought of freelancing, if for no other reason than to learn how to do it to have workable back-up plan in place?

My thought process for posing these questions comes from the fact that freelance writing has been my back-up employment plan since I retired from the military. In reality, it's my long-term employment plan not just a contingency.

The great thing about being a writer is that once you are taught how to put together articles and stories—or have taught yourself on your own—you realize you have a special skill that has many uses and is not limited to any particular job or location, especially if there's an Internet connection nearby.

I prefer the lifestyle of a freelancer. If I were still a reporter or any sort of staff writer, in my spare time I would build an edocument collection for a rainy day. If nothing else, it would place me ahead of the power curve while observing happenings in the publishing world.

17

Nonfiction Book Proposals

o o o o o o o o o o o o o o o o o o o o o o o o o o o o
A hefty, overblown production whose insane initial price ensured its arrival in the remainder shops by the direct route from the warehouse.

—*Clive James, 1939*–

I've put together a number of nonfiction book propos-als over the years. I'm certain with time I could con-vince a publisher to go to print with book ideas. I don't see this activity as being particularly hard. I do see it as being time-consuming.

I like to think of myself as the most persistent per-son on the planet. I have an impeccable work ethic when I'm working. I can also be a professional goof-off if I'm in the mood. I like to believe I'm a decent writer.

I don't consider this overconfidence. I believe in power inherent to positive attitudes.

Meanwhile, I feel that I must eventually make a living as a writer. To spend time beating around the bush with book ideas, as well as exposing material that could be borrowed by others—ideas are not copyrightable—does not sound like an efficient pursuit. I'm not implying I believe someone would steal my book ideas. However, at the same time I would prefer to receive as much compensation as possible from my creativity and any work involved.

If I were to sit down to discuss why I began selling nonfiction text in edocument format, the first thing I would point out would be the professional writers who are selling what essentially amounts to nonfiction book chapters as online edocuments. They don't appear to have wasted time with book proposals, table of contents, front covers, front and back matter, or indices.

They posted their writing to the Web and readers are buying it.

18

Internet Efficiency 101

○ ○

Someone told me that each equation I included in the book would halve the sales.

—Stephen William Hawking, 1942–

I've experimented with affiliate marketing programs since I began penning Web content. One factor I point out to affiliate marketing newcomers is the math enjoyed by a portion of those involved. The reason affiliate marketing—a combination of hyperlinks, advertising copy and Web graphics—works so well—the reason there are so many advertising dollars changing hands—is because of the Internet's ability to channel consumers through Web sites, tracking sales and paying percentages along the way.

Here's a simple description of the math involved with affiliate marketing:

- An online store sells kites, for example. The store has done well since the inception of online sales, but would like to expand.

- The marketing department attends a sales conference and becomes convinced that affiliate marketing will expand their customer base and subsequent online sales by way of enticing other Web sites to advertise their kites for a percentage of sales.

- They open their Web site affiliate program, begin offering it to the world, and are immediately inundated with affiliate marketers who see the revenue potential. Everyone gets busy and soon the online kite store experiences an astronomical jump in sales.

- They are now working with eighty thousand affiliates—other Web site owners—that advertise their various kites. The affiliates post banner ads and pen textual ads, and jam their sites into search engines with whatever means necessary to ensure that their sites show up in top search results.

At this point, it doesn't matter what percentage the kite company or affiliates receive because of the math involved:

- Every time an affiliate makes a sale the kite company pays the affiliate 50¢ and the kite company also receives 50¢ (in profit).

- With eighty thousand affiliates selling one kite a month, the kite company receives $40,000 a month (in profit) and each affiliate receives 50¢.

This is the aforementioned "consumer channeling" effect. Now we'll turn up the volume on the math:

- A different online kite store—a bigger one—has a million affiliates and all affiliates are selling a hundred kites a month. The kites are the expensive variety and the affiliates receive $10 every time they sell a kite.

- The affiliates are now generating $1,000 a month for their personal freelance coffers—probably not enough for a copywriter to subsist—but better than 50¢ a month.

- If the store, like the affiliates, receives $10 per kite sale, how much money did the store generate in profit in a month's time while channeling? My little plastic calculator tells me the online store pulled in $1 billion in a month's time.

The reason you should understand consumer channeling is because the only Web sites that generate respectable revenue from affiliate marketing do so by channeling—whether a mainstream Web entity that entices every Web site on the planet to advertise on every other Web site on the planet and vice versa—or an affiliate attempting to subsist as a freelance copywriter with one or more small Web sites.

How did my affiliate marketing experiences affect my desire to publish edocuments? The phenomenal success of some edocument publishers could be attributed to the fact that one writer is selling to a captive, worldwide audience, which is very similar to channeling if you stop to think about it.

On that note, I should point out that the E-Doc-S Web pages http://www.e-doc-s.info that support this book project do not contain affiliate marketing hyperlinks because I know such links are often unappreciated by other affiliate marketers.

On the other hand, note that edocuments at E-Doc-S *do* contain affiliate marketing hyperlinks, and said thought process caused me to word edocument notice of copyright like so:

> After purchase, you are free to e-mail the edocument as often as you like to multiple addresses. You may also use it in digital or paper format within a non-profit or public domain teaching environment. You may not alter text, illustrations, charts, hyperlinks, or other content in any way, shape or form. The document must remain in its original digital or paper state to be reused. You may, of course, expand on any ideas posed within a case study (that's the purpose of a case study). If you quote (fair use) an excerpt or refer to a chart, please note the case study number and author (by name, including quotation marks around names that have quotation marks; some of our authors

use pen names). No portion of an edocument may be reprinted or reproduced for commercial gain.

The above verbiage may be somewhat arguable. Feel free to register your thoughts at our Gather location, if you like. Note that it is original thought process based on pragmatism, and nothing else. If you think there's a comparison to be made between the copyright notice wording and the creative commons notices widely available on the Web, please also feel free to do that.

19

Internet Efficiency 102

○ ○

How much wood would a woodchuck chuck, if a woodchuck could chuck wood?

—Rebus Rhyme

Google made two relatively recent changes at Blogger. One you may have heard of: they *bought* Blogger. The other one you may not have heard of, although it is a huge perk for modern-day freelance writers.

Before I continue, the connection that exists between edocuments and blogging is *efficiency inherent to the Internet*: Blogging carries with it the potential to eliminate newspapers, syndication agencies, advertising departments and a number of editorial positions.

After Google bought Blogger they placed it in beta mode to make changes. When they reopened for business, they had revised the Blogger dashboard code so that the dashboard replicated the function of an RSS reader. If you've never heard of RSS, there are many ways to describe what it does, but the best description is to state that RSS is a Web-based form of syndication.

In short, if you build blogs using Blogger, you can now log on to one of your blogs—or log on to any Google account (then return to Blogger)—and your Blogger Dashboard will display links to your separate Blogger blogs on a single page.

In so far as generating freelance writing revenue via blogging with a number of blogs, if you use AdSense advertisements within your blogs, you may reach the point where you have to ask yourself: "How much content and subsequent AdSense revenue am I capable of generating as a writer?"

If you've also never heard of AdSense ads, Google essentially developed Web advertisements that use "artificial intelligence" to display content relevant to Web pages on which they appear.

If you became bored with blogging over time, you should reconsider by Googling the phrase "Amazon blog cares for lunch" to observe the book link that appears. If you decide to start blogging again, create ten blogs this time, and manage them using the new Blogger dashboard.

Interestingly, the original edocument I wrote covering blogging as a way to generate freelance revenue was originally seven pages. After Google created the new Blogger dashboard RSS function, the edocument shortened to one page. As I continue to tweak the idea, it's now down to about two paragraphs.

Sorry for plugging a specific edocument within the pages of this book. I promised myself I wouldn't do that in the first edition.

20

Advertising

The following is a short list of advertising innovations that utilize power inherent to the Internet, as well as electronic publishing, and publishing in general. Each method details a way to advertise or promote portions of an edocument collection (or the whole thing) by way of allowing advertising to advertise for itself.

• The aforementioned innovative copyright notice could be used as a means to encourage readers to send your edocuments to other people. Ensure that you have hyperlinks in your edocuments that return readers to your main edocument sales platform. The idea for pointing this out is to open your mind to viral advertising methods that may be obtained from edocuments.

- Ensure you are "credentialed" as much as possible. There is no requirement to have a university degree to be a writer, but you need to be a subject matter expert on something, and a good writer. Your edocument sales platform should outline your wealth of knowledge, branching out from your main page with hyperlinks to your various edocuments. There is nothing more powerful than an expert—self-taught or otherwise—who is capable of illustrating with words. Also establish a certain number of edocuments you give away for free.

- Write a nonfiction book and note the link to your edocument sales platform in the back matter. Advertise individual edocuments in back matter as well. Ensure that any books that support your Web or edocument efforts are advertised within your edocuments and on your Web pages.

- Most Web design packages come with what essentially amounts to free Web advertising in the form of meta tags. Internet users searching for words or phrases will arrive at your site if your meta tags are correctly established.

- Build a social networking community discussion area with links to your edocuments similar to the one I set up at Gather for this book <u>http://e-doc-s.gather.com</u>.

Internet efficiency allows much more bang for your advertising buck, if you have to spend any money at all. To prove my point, call your local newspaper to ask what it costs to advertise a nonfiction book over a period of time in the paper.

An expanded version of the above list is available in the Gather discussion area.

21

Protect Your Edocuments—Protect the World

Viruses may be a concern with edocument publishing, but I don't think they are a serious concern. It was not too long ago that e-mail attachments were scary things. Now, Web e-mail providers scan attachments before they become part of the provider's servers, and the attachments are as ubiquitous as e-mail itself. Having said that, I still don't open email attachments unless I know the sender.

Moreover, that doesn't mean freelance writers should not be extra careful. We definitely should not be perpetuating computer security problems by operating in unsafe computer environments.

Here's a solution I've been tweaking for nearly as long as I've been researching edocument sales. Assumption is that at least a few writers concurrently use computers for writing and Internet access:

- If you do not have an *administrator* user account with user name and password established on your home or work computer, you should create one immediately.

- If you do not have a *second* user account with user name and password established, you should create one of these too.

- If you stay logged on to your administrator account for lengthy periods of time while surfing the Internet you may be a computer security disaster waiting to happen.

The easiest way for a virus to enter your computer is through your administrator account, and many viruses and security breeches are extremely hard to delete or quarantine.

Obviously, you have to be logged on as administrator to download software and security updates from the Internet for your operating system. The trick is to stay logged on for as short a time as possible.

Three computer security issues to consider as freelance writers (in order of precedence):

- If you lose a computer hard drive to a virus, you can't work; unless you're adept with a typewriter.

- Viruses have a number of purposes. If your administrator account and second user accounts are not secure with user names and passwords, it's possible an automated virus could enter your computer to copy your hard drive. Means your writing projects could end up anywhere.

- If you plan to create an edocument collection, you should take extra steps to ensure your computers and documents are secure to preclude inadvertently loading virus-laden edocuments to mainstream servers.

Basic security precautions that will assist you with having less computer security problems are as follows:

- If you would like to become overly protective with your computer security, ensure that you use at least eight characters for your administrator user name and password *and* second-user user name and password. Use a combination of letters, numbers, and symbols (e.g., COMPUTERstrong5678@#$). Keep in mind that user names and passwords are case sensitive. Legibly write down all user name and password information to store in a safe location.

If you lose the information you may have to reformat your hard drive to gain access to your computer.

• Do not use the words "administrator" or "admin" as an administrator or second-user user name.

• Do not use the word "password" or "pass" as an administrator or second-user password.

• Turn off all other user accounts.

• If you use a computer network, ask your network administrator to ensure that the above is accomplished for you. Your network administrator may have assumed you knew to establish these accounts on a laptop computer that you periodically connect to the network (for example).

• Ensure your operating system firewall is switched to the *on* position.

• If you have anti-virus software that establishes a separate firewall in the *on* position that's OK too. One of the two firewalls must be *on*.

• Firewall all individual computer connections that are not automatically firewalled by your main firewall.

• Purchase or free-download as much anti-virus software as you can possibly load on your computer(s) without causing internal conflicts.

• Purchase or free-download as much anti-*everything* software as you can possibly load on your computer(s). Anti-everything software includes protection against unwanted advertisements, software that spies on your computer, identity theft and any other privacy issues you can think of.

- The E-Doc-S Gather location <u>http://e-doc-s.gather.com</u> lists free anti-virus and anti-everything software.

- Set your computer operating system to *auto-update*.

- Run manual anti-virus scans often and establish scheduled scans to run when your computer is on but not in use.

In a nutshell, it is impossible to stop a determined, automated virus from entering your computer, but as noted above there are additional layers of protection you may add to your computer to minimize risk. The general idea is to force viruses to skip past your computer as they search for places to do damage.

22

Work Ethic

o o o o o o o o o o o o o o o o o o o o o o o o o o o o o o

Anyone could write a novel given six weeks, pen paper, and no telephone or wife.

—Evelyn Waugh, 1903-66

In my experience, all writers seem to have different work habits. I suspect many of my peers stay up working late into the night. I know some of my friends do. Whereas, I am the proverbial early bird—fully capable of rousting myself between zero-two-hundred and zero-four-hundred for days on end—especially when immersed in a project. For me, starting my day after watching begin-morning-nautical-twilight is powerful stuff, but again, everyone is different and I'm sure lots of writers watch BMNT before retiring for the day.

(Sorry for the milspeak. I've always wanted to include a reference to "BMNT" in nonfiction text. BMNT is a military term that designates the specific time when the heavens begin to lighten in the morning a half an hour or so before sunrise. It's probably more of a tactical reference than anything else. In the military, if you're not awake by BMNT, someone might sneak up on you and … you know.)

The point for discussing work hours is because I believe edocument publishing may offer a number of solutions for problems often associated with taking the full-time freelance writer plunge:

- I believe there may be more money in edocument publishing than with traditional fee-based freelance writing, which in turn means an edocument collection could be used to establish a stream of income to allow a writer to take the plunge with less risk. Quitting a day job can give anyone the jitters.

- There is less credentialing required, since the normal publishing Catch 22 that states "you must be published before you can publish" no longer applies. A writer must have some sort of professional background to create an online edocument sales platform, but there's no bureaucratic or editorial process for saying "no" to a determined writer.

- Once you have a solid collection of edocuments supplying you with royalties, you may have more time available during the day to do things other than write. I like to travel the world in my off time, but that's just me. Maybe you like to hang out at the pool. A crafty writer could use revenue generated from edocument sales to create needed where-withal and time to begin a fiction career.

• If you combine the thought process of the above three ideas, edocument publishing may offer an out for writers who are stuck trying to start or kick-start a writing career with a poor writing environment on their hands. It only takes an hour or two a day to get started with an edocument sales campaign. After that it's just a matter of persistence.

With traditional freelance writing I found I needed hours, days, weeks, and months of undisturbed writing time. For some people that's just not possible.

The remarkable thing about an edocument publish-ing platform is that it's real writing, in that, everything that pertains to traditional nonfiction writing also applies to edocument publishing. As the market opens up, writers will either know from experience or discover by trial and error that the same moral, ethical, and legal rules apply. I'm not sure what the appropriate term would be for people incapable of dreaming up and crafting their own creative projects, but I definitely wouldn't call them "writers."

Less work and more pay sounds like a heck of a good deal to me and I wouldn't want to toy with that proposition.

Interactive Resources

To observe a functional edocument sales platform and sample edocuments, visit http://www.e-doc-s.info.

All questions, suggestions or other feedback concerning this book should be noted at http://e-doc-s.gather.com or in a book review at http://www.amazon.com.

General Internet Resources

A

Adobe Reader and Acrobat software may be found at http://www.adobe.com.

Amazon http://www.amazon.com: The most amazing Web site on the planet.

Amazon Advantage Program http://advantage.amazon.com: The world's first and best affiliate program.

Amazon Associates Program http://associates.amazon.com: In-road to Amazon sales for self-publishers.

B

Brown Herron Publishing http://www.brownherron.com: My heroes.

C

ClickBank http://www.clickbank.com: Option for accepting credit card sales.

E

E-Doc-S http://www.e-doc-s.info: Main Web site that supports this book.

E-Junkie http://www.e-junkie.com: My favorite digital sales Web site.

G

Gather discussion area http://e-doc-s.gather.com: Feel free to join in.

Google Base http://base.google.com: Awaiting many good things from our friends at Google.

Google Checkout http://checkout.google.com: Check 'em out.

Google Docs & Spreadsheets http://docs.google.com: Thank you.

Google Page Creator http://pages.google.com: Expect many good things to evolve from this software.

K

Kinn, Sean: A complete list of Sean Kinn's works may be found at http://www.seankinn.info.

L

Lulu http://www.lulu.com: Coolest eBook sales site on the Web.

M

Microsoft Office Live http://officelive.microsoft.com: Highly recommended for freelance Web design.

P

PayPal http://www.paypal.com: Ubiquitous option to accept credit card sales.

Y

Yahoo! SiteBuilder http://sitebuilder.yahoo.com: Expect many good things to evolve from this software.

2

2Checkout http://www.2checkout.com: Option for accepting credit card sales.

Afterword

This book is small; ten thousand or so words. I kept it short to get the information on the street as quickly as possible. I considered stretching it out to fifty thousand words, but said "nah" to myself and nipped the thought process in the bud. It would have taken too long and much of the text would have ended up fluff.

If it's not obvious, I'm attempting to give the edocument market a push because I think this new form of writing—for lack of a better term—is cool.

If you're asking yourself why someone would write a book about edocument publishing processes, from my perspective I think books printed on paper are an extension of the information search process as a whole. In other words, if you ran across this little book in a bookstore … well, you see my point. The Internet has changed a lot of things in our lives, but it's really just another form of communication.

If nothing else, term this book a "news release." The next edition will probably expand a bit but not much.

978-0-595-45135-7
0-595-45135-7